IMPROVISED BATTERIES and DETONATING DEVICES

DESERT PUBLICATIONS

IMPROVISED BATTERIES
and
DETONATING DEVICES

ISBN: 0 - 87947 - 219-7

DESERT PUBLICATIONS

Cornville, Arizona 86325

Table Of Contents

Tables

Illustrations

Abstract

BATTERY

A discussion of battery fundamentals is presented to familiarize the reader with the nomenclature and conventions used in discussing the batteries and other devices investigated.

Although other factors were considered, two criteria weighed most heavily on the selection of indigenous batteries: availability of materials and current output. Batteries made of the more generally indigenous materials, urine for example, were not able to furnish sufficient current. The final test of the successful battery was the detonation of the No. 6 blasting cap.

Two batteries are recommended. The lye battery is the most powerful but has poor endurance. The ammonium chloride battery, if properly enclosed, can be used ten hours or more after it has been activated.

Remote activation of the lye battery could not be accomplished. Satisfactory results were obtained with the high endurance battery and a switch which can be closed by foot or vehicle pressure.

In addition to the recommended batteries, a six volt, three watt bicycle generator and a six volt automobile generator were successful in detonating the No. 6 blasting cap.

TRIGGER DEVICE

A simple, inexpensive electronic trigger device has been designed, fabricated, and tested for use in demolition, booby trap, and perimeter defense applications. The interruption of a minute current caused when a fine trip wire is broken is used to switch an electric current to a detonator, claymore mine, alarm or similar device. Tests conducted to date show its feasibility. Recommendations for additional work include final selection of operation current and voltage as determined by user requirements, and tests to assure conformance with environmental specifications.

GELS

Oil gellation in automobile engines causes permanent damage to moving parts due to oil starvation. The most successful gelling agent among those tested is N-coco-γ-hydroxybutyramide. This substance can be added to the engine oil presynthesized, or it can be formed *in situ* by the reaction between coco amine and butyrolactone.

A polyacrylamide gelling agent was used to completely gel an automobile engine cooling system.

PERSONNEL MARKERS

Several chemicals were investigated as personnel markers. The fluorescent dyes and powders were the most effective in this application.

Section I.
Indigenous Batteries

I. Battery Fundamentals

The following brief discussion of battery fundamentals is offered to establish the nomenclature and conventions to be used in the text of this report.

Chemical reactions which cause a shift of electrons from one chemical species to another are called oxidation-reduction reactions. Oxidation is the process of electron loss, and reduction that of electron gain.

Whenever the oxidation half-reaction is separated from the reduction by a conducting medium, a potential difference occurs. The area at which the oxidation half-reaction occurs is called the anode. The place where the reduction half-reaction occurs is the cathode. The cathode and anode are called electrodes. The unit of potential difference is the volt. One cathode-anode pair is a cell. A battery is two or more cells connected together.

The maximum potential difference, or electromotive force, of a cell is a fundamental characteristic of the chemical reaction occuring in the cell. The maximum potential difference is a measure of the driving force behind a chemical reaction in that it indicates the extent to which the combined reactants deviate from the situation of chemical equilibrium. Large values of potential difference are usually caused by rapid, irreversible chemical changes.

Unless the cathode and anode are connected by an electrical conductor, no electrons can flow and the oxidation-reduction cannot proceed. Theoretically, a cell can maintain its energy indefinitely in this situation, but does not because of competing chemical reactions which occur in spite of the fact that no electrons can flow in the circuit. These reactions expand chemicals and coat the electrodes. When the cathode and anode are connected by a copper wire the electrons can flow and the oxidation-reduction can proceed.

II. Criteria for Indigenous Batteries

The following factors were established to measure the qualifications of the batteries investigated.

A. Availability of Materials.
B. Current and Potential Difference.
C. Endurance.
D. Bulk and Ease of Construction.

A. AVAILABILITY OF MATERIALS

The choice of materials for an indigenous battery depends to a great degree upon the area of the world in which the battery is to be built. There are, however, materials which do not depend on geography or culture: for example, human secretions and charcoal. We considered materials of this type most desirable for indigenous batteries, providing they furnished adequate current and potential difference.

Slightly less useful materials are those which are manufactured in many underdeveloped areas. These materials are to some extent dependent on locale. Examples are lye and many metals.

Materials which can be obtained in stores, houses and on farms throughout much of the world make up a third and least indigenous class of materials. Examples in this class are soldering fluxes, chemical fertilizers, disinfectants and the interiors of dead batteries.

B. CURRENT AND POTENTIAL DIFFERENCE

Deficient potential difference in a cell was not, by iteself, considered critical, since this can be overcome by making a battery of two or more cells in series. Low current (high internal resistance), however, is taken as sufficient reason for rejecting a cell, since overcoming inadequate current requires a relatively large number of cells. For example, in order to obtain a battery furnishing one ampere current and one volt from cells individually furnishing 0.7 volt and 0.2 ampere, a minimum arrangement of four pairs of parallel sub-batteries, each consisting of two cells in series, would be required. This is a total of eight cells.

Cells which furnished a no load current in excess of one ampere were tested for their ability to detonate a No. 6 blasting cap. The firing circuit consisted of the cell or battery, the ohmic equivalent of three hundred feet of No. 18 gauge copper wire and one or more blasting caps. The No. 6 has a bridge resistance of 1 ohm and a recommended all-fire current of one ampere. It was necessary that a battery fire the No. 6 blasting cap before being recommended as an indigenous battery.

C. ENDURANCE

A durable energy source is necessary for the operation of delayed action devices such as mines and booby traps. For this reason, one of the recommended batteries has high endurance.

D. BULK AND EASE OF CONSTRUCTION

As mentioned above, low current cells have not been recommended because of the large number of cells necessary to correct deficient current. It is also in the interest of compactness that semi-dry cells are recommended over wet cells.

On the other hand, compactness has its drawbacks. A cell with a small electrode area has high internal resistance and low current. Small cells are also difficult to construct and handle, expecially when corrosive chemicals are used.

Our most satisfactory results have been with cells about two to three inches square and one-fourth to one-eighth inch thick. These dimensions furnish a compromise between compactness and operational requirements.

III. Individual Indigenous Batteries

The batteries investigated under this contract fall generally into four areas:

A. Organic Batteries - those made from human secretions, plant extracts, etc.

B. Lye Batteries.

C. Ammonium Chloride Batteries.

D. Miscellaneous Batteries.

A. ORGANIC BATTERIES

The organic batteries that were investigated are listed, along with their no-load potential difference and current, in Table I.

The urea acetate battery gave the best results of the organic batteries. Urea acetate refers to a solution of vinegar and urine which has been concentrated by boiling to about a tenth of its original volume and which contains some urea diacetate.

$$\underset{\text{urea}}{NH_2CONH_2} + \underset{\text{acetic acid}}{2CH_3CO_2H} \rightarrow \underset{\text{urea diacetate}}{CH_3OCONH_3CONH_3OCOCH_3}$$

It was reasoned that the solution of a salt of a weak acid (vinegar) and a weak base (urea) would have better conducting properties than either the acid or the base. This view was substantiated by the higher current values obtained with this battery. Current values proved unreliable, however, and depended to some extent upon the source of urine.

Attempts to detonate the No. 6 blasting cap with a ten cell urea acetate battery failed. Although no load current values seem relatively high, when a load is placed in the circuit the current rapidly dissipates to very low values.

Sodium chloride was added to the urea acetate cells and

TABLE I
ORGANIC CELLS

Battery	No Load Potential (Volts)	No Load Current (Amperes)	Comments
Al-Urea Acetate-X-C-Cu*	0.5	0.2	Ten cell battery fails to detonate the No. 6 blasting cap
Al-Urea Acetate-X-C-Cu	0.6	0.1	
Al-Urea Acetate-NaCl-C-Cu	0.6	0.2	Ten cell battery fails to detonate the No. 6 blasting cap
Al-Urea Acetate-NaCl-Mud-Cu	0.8	0.06	
Al-Vinegar-NaCl-X-Cu	0.75	0.1	Liquid Cell
Al-Coca Cola-X-X-Cu	0.5	<0.01	Liquid Cell-gas evolved
Al-Hydrolyzed Tea-X-X-Cu	0.5	<0.01	Liquid Cell-tea hydrolized with washing soda (Na_2CO_3)
Al-Saliva-X-C-Cu	0.25	<0.01	
Al-Saliva-NaCl-C-Cu	0.35	<0.01	
Al-(Saliva-Starch)-NaCl-C-Cu	0.25	0.09	
Al-(Saliva-Starch)-C-Cu	0.85	<0.01	
Al-(Saliva-Starch)-NaCl-C-Cu	0.65	0.1	

*Electrode Aluminum (Al) — Electrolyte Urea Acetate — Depolarizer None (X) — Carrier Charcoal (C) — Electrode Copper (Cu)

some of the other organic cells to enhance current flow. Very little or no improvement was experienced as a result of this addition.

The hydrolized tea cell was made from tea that had been boiled with washing soda (sodium carbonate.) The product contained sodium salts of the complex phenols, called tanins, found in tea. It was reasoned that the sodium salts might be more conductive than the untreated tea. No results were obtained to indicate this was the case.

The saliva-starch battery was built to test if the action of the enzymes in saliva on starch at an electrode might produce a satisfactory cell. Saliva-starch cells exhibited very low current.

The organic cells all have high resistance and low current flow. For this reason none of these cells have been recommended as indigenous batteries.

B. LYE BATTERIES

The most generally acceptable cell among those tested is the lye or sodium or potassium hydroxide cell. Although short lived, this cell is the only single cell capable of detonating the No. 6 blasting cap. The various types of lye cells that have been constructed are listed in Table II with current and potential difference data.

Lye, which is sodium or potassium hydroxide, is produced from wood ashes for use in the primitive manufacture of soap. Solid lye can be purchased in grocery stores or it can be obtained from drain cleaners such as "Draino."

Satisfactory results are obtained only with concentrated solutions of lye. Figure 1 illustrates the dependence of current on lye concentration. Concentrated solutions of lye can be obtained by dissolving in a given weight of water an equal weight of solid lye. Dilute solutions of lye can be concentrated by boiling.

The recommended lye cell is a semi-dry cell. Semi-dry cells use an adsorptive substance or filler to hold the battery chemicals between the electrodes. Pulverized wood charcoal is easily obtained and works very well as filler material. The

TABLE II
LYE CELLS

Battery	No Load Potential (Volts)	No Load Current (Amperes)	Comments
Zn-NaOH-MnO_2-C-Cu*	0.9	0.2	
Zn-NaOH-MnO_2-C-Cu	1.0	0.2	
Zn-NaOH-$KMnO_4$-C-Cu	1.85	1.5	
Zn-NaOH-$KMnO_4$-C-Cu	1.85	1.2	
Zn-NaOH-$Ca(OCl)_2$-C-Cu	1.75	1.6	$Ca(OCl)_2$ is Perchloron commercial bleach and disinfectant
Zn-NaOH-$Ca(OCl)_2$-C-Cu	1.65	0.4	
Zn-NaOH-Ca $(OCl)_2$-C-Cu	violent reaction		Tin was from tin can
Zn-NaOH-$KMnO_4$-C-Cu	0.25	0.1	Tin was from tin can
Zn-NaOH($KMnO_4$-S)-C-Cu	violent reaction		
Zn-NaOH-$Ca(OCl)_2$-C-Cu	violent reaction		Commercial activated charcoal was used
Zn-NaOH-$KMnO_4$-Mud-Cu	0.5	0.2	
Zn-NaOH-NaOCl-C-Cu	0.5	0.1	NaOCl is "Clorox" bleach
Zn-NaOH-NaOCl-Cu	1.9	0.2	Liquid cell

*Electrode Zinc (Zn) — Electrolyte Lye (NaOH) — Depolarizer Manganese Dioxide (MnO_2) — Carrier Charcoal (C) — Electrode Copper (Cu)

pulverized nest of the mud-dauber wasp is also satisfactory. Plain mud does not work well, however, and neither does cardboard.

Zinc and galvanized iron have proved to be the best reacting electrodes although aluminum is also satisfactory. Tin did not work well. This cell is short lived because the lye reacts with the zinc, even though current is not permitted to flow.

To obtain maximum current with a lye cell, a depolarizer is necessary. Chemicals which serve satisfactorily in this capacity are calcium hypochlorite ("Perchloron", "HTH", "Pittchlor", "Hychlor"), a bleach and disinfectant, and potassium permanganate, a disinfectant. Both of these compounds are dangerous and should be kept out of contact with skin and clothes: Sodium hypochlorite ("Clorox") did not enhance current flow.

A properly constructed single lye cell will reliably detonate a No. 6 blasting cap in the circuit described on page eight. Two cells will detonate two No. 6 blasting caps in series or in parallel.

To construct the lye battery, moistened filler material and depolarizer are placed between the two electrodes. The cell can be transported in this condition without deterioration. When it is time to use the battery, the zinc or galvanized iron electrode is removed, lye solution is poured over the filler and depolarizer, and the electrode is replaced. Leads from the detonators must be connected to the cell within the next two or three minutes.

Several experiments were conducted with the lye battery to determine the feasibility of a remote activation device. A plastic bag filled with lye solution was connected to the cell. A board was placed over the bag. When the board was stepped upon, the bag was compressed and lye was spewed into the cell. No blasting cap detonations were obtained using this procedure. Apparently, a load on the cell at the time of activation makes a sufficient current surge impossible. Satisfactory results were obtained, however, using a more durable battery and a switch can be closed by foot or vehicle pressure.

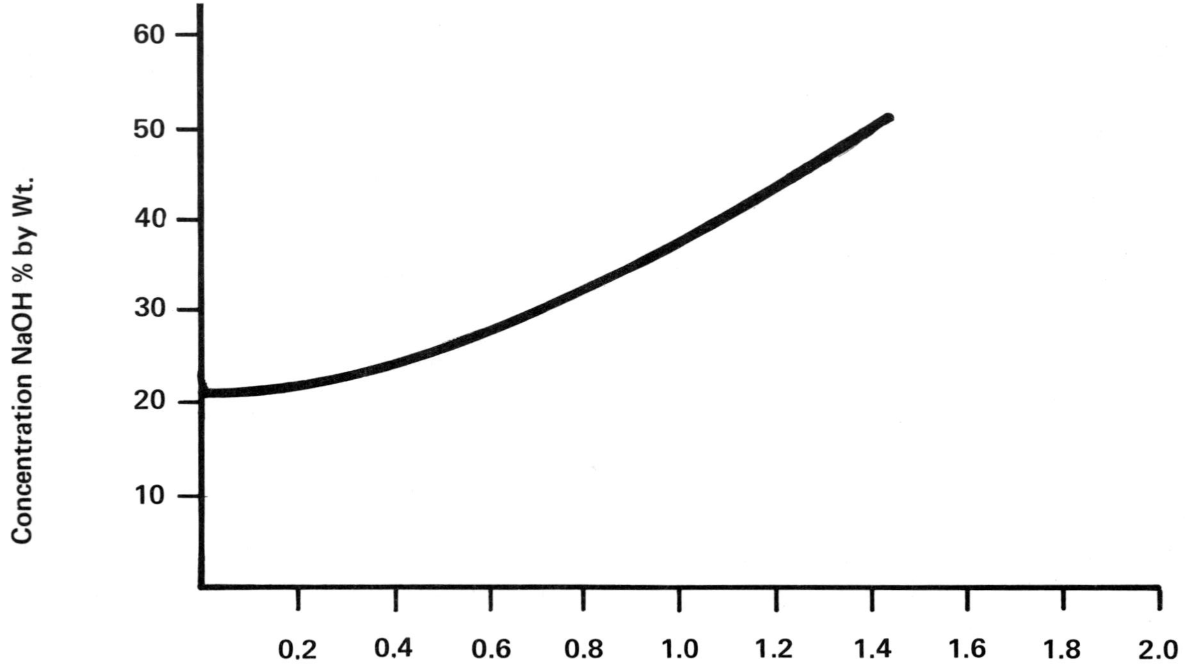

FIGURE 1
Current Produced by the Lye Cell vs Lye Concentration

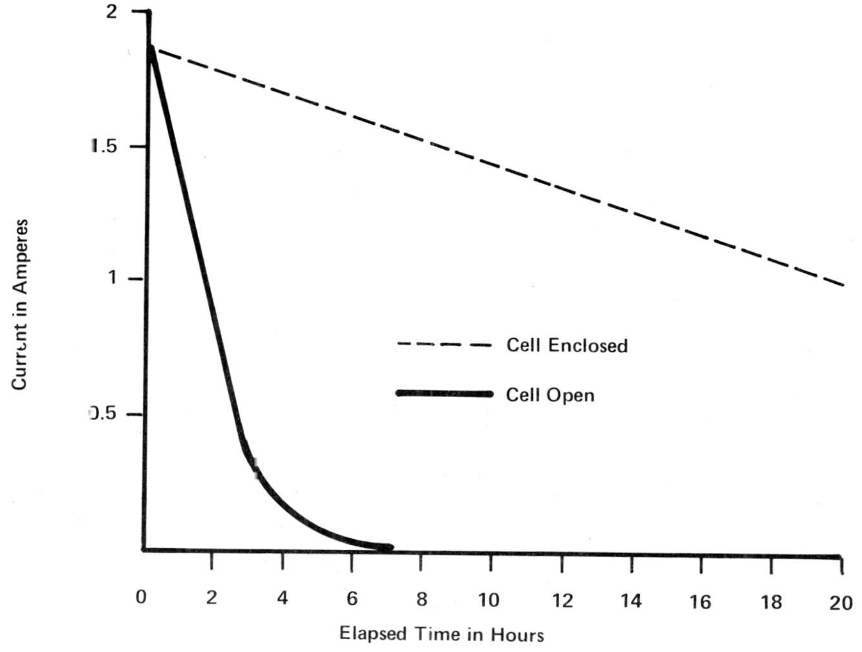

FIGURE 2
Endurance of the Cell Zn-NH_4Cl-Dead Dry Cell Interior-C-Cu

TABLE III
AMMONIUM CHLORIDE CELLS

Battery	No Load Potential Difference (Volts)	No Load Current (Amperes)	Comments
Zn-NHCl-X-Cardboard-Cu*	0	0	Ten of these cells, five series pairs connected in parallel, detonated thr No. 6 blasting cap
Zn-NH_4Cl-X-C-Cu	0.5	0.2	
Zn-NH_4Cl-MnO_2-C-Cu	0.84	1.4	
Zn-NH_4Cl-Dead dry cell interior-C-Cu	0.93	1.8	
Zn-NH_4Cl-Dead dry cell interior-C-Cu	0.5	2.0	
Zn-NH_4Cl-NH_4NO_3-C-Cu	0.6	0.3	
Zn-NH_4Cl-$KMnO_4$-C-Cu	0.8	2.4	
Zn-NH_4Cl-$NaNO_3$-C-Cu	0.75	1.0	
Zn-NH_4Cl-$Ca(OCl)_2$-C-Cu	Explosive!		
Zn-NH_4Cl-Dead dry cell interior-C-Cu	0.8	0.3	Copper electrode copper wire wrapped around a board.

*Electrode Zinc (Zn) — Electrolyte Ammonium Chloride (NH_4Cl) — Depolarizer None (X) — Carrier Cardboard — Electrode Copper (Cu)

C. AMMONIUM CHLORIDE BATTERIES

The advantage of the ammonium chloride cell lies in its endurance. Whereas the lye cell had to be used shortly after activation, the ammonium chloride cell can be used hours after it has been activated, providing it is protected against loss of moisture. Figure 2 illustrates the endurance of the cell Zn-NH_4Cl-dead dry cell interior-C-Cu, both when sealed against loss of moisture and when left open to the atmosphere.

Ammonium Chloride can be obtained in most areas of the world under the name sal ammoniac. It is also present in some solid fluxes, and "tinning blocks" used in soldering sheet metal. In order to obtain high current values, the ammonium chloride cell requires a depolarizer. Manganese dioxide (pyrolucite) from dead dry batteries serves well in this capacity. Potassium permanganate is also a good depolarizer for the ammonium chloride cell. The calcium hypochlorite depolarizer used in the lye cell reacts explosively with ammonium chloride.

Zinc, galvanized iron, or aluminum make satisfactory reacting electrodes for the ammonium chloride battery, although the galvanized iron deteriorates once the zinc veneer is used up. The metal must be clean for best results. A copper electrode made by wrapping copper wire around a board resulted in a cell with insufficient current.

The possibility of rejuvenating dead dry cells by adding ammonium chloride through holes drilled in the cell casing was investigated. This procedure proved unreliable. When a dry cell has been completely used up, the zinc electrode is usually so badly deteriorated as to make rejuvenation impossible. We found that the more dependable procedure was to remove the chemicals from the battery interior, add additional ammonium chloride and use fresh zinc for the electrode. The ammonium chloride cells investigated in this program are listed in Table III.

The ammonium chloride battery made of three cells connected in series is capable of exploding one No. 6 blasting cap in the circuit described on page eight. A battery consisting of five cells will detonate two blasting caps.

The ammonium chloride battery is constructed in its en-

TABLE IV
MISCELLANEOUS CELLS

Battery	No Load Potential Difference (Volts)	No Load Current (Amperes)	Comments
Zn-NaCl-X-X-Cu*	0.6	$<$ 0.01	Liquid Cell
Zn-NaCl-X-C-Cu	0.7	0.06	
Zn-NaCl-X-Mud dauber nest-Cu	0.75	0.2	
Zn-NaCl-X-Mud-Cu	0.75	0.06	
Zn-Na_2CO_3-Cu	0.65	$<$ 0.01	Liquid Cell
Zn-Na_2CO_3-Cu	1.15	$<$ 0.01	Liquid Cell

*Electrode	Electrolyte	Depolarizer	Carrier	Electrode
Zinc (Zn)	Ammonium Chloride (NH C1)	None (X)	Cardboard	Copper (Cu)

tirety prior to use. It should be placed in a closed container or wrapped in some material such as waxed paper to prevent loss of moisture. When wrapping the battery, one must take care to prevent chemicals from coating the inside of the paper and creating a short circuit between the cells.

A battery, made of ten ammonium chloride cells from which the depolarizer was omitted was able to detonate the No. 6 blasting cap. This battery consisted of pairs of cells, one cell connected to the other in series. Each pair of series cells is connected to the other pairs in parallel.

D. MISCELLANEOUS BATTERIES

Table IV lists some additional cells that were investigated. The cell Zn-NaCl-X-mud dauber's nest-Cu shows surprisingly high current.

IV. Generators and Alternators as Indigenous Sources of Electrical Power

In most parts of the world today, bicycles and even automobiles are becoming commonplace. Bicycle and automobile generators will successfully detonate the No. 6 blasting cap in the circuit described on page eight.

Generators possess the advantage of endurance over batteries. What is more, they can be operated by simple spring devices triggered by snares.

Our experiments were performed with Japanese, French, and English made bicycle generators rated at six volts and three watts. When the drive wheel of the bicycle generator is rolled sharply across the palm of the hand, enough electric power is created to detonate at least one No. 6 blasting cap in the circuit described on page eight. The Japanese-made generator detonated two caps in series.

A six volt automobile generator with the field coil jumpered to the armature has also been successfully used to fire the No. 6 blasting cap. The generator must be operated by pulling a cord wrapped around the generator drive pulley.

A twelve volt automobile alternator, operated by means of a cord wrapped around the drive pulley, could not be made to detonate the No. 6 blasting cap.

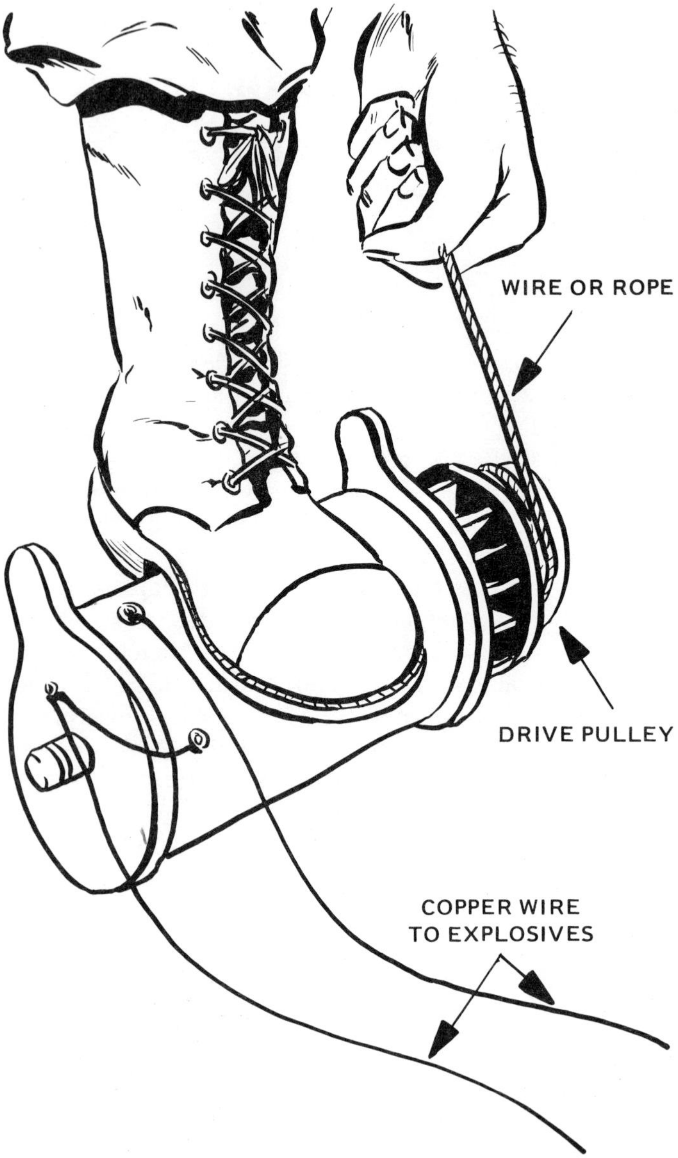

FIGURE 3
Using An Automobile Generator To Set Off Explosives

USING AN AUTOMOBILE GENERATOR TO SET OFF EXPLOSIVES

An automobile generator when properly wired, will set off one No. 6 blasting cap or one S95 squib.

MATERIALS

Automobile generator (6 volt)
Copper wire

PROCEDURE

1 An automobile generator has three terminals. They are usually labeled G, F, and A. The G terminal may also be labeled C. Connect the A and F terminals together with a piece of copper wire. Scrape the wire at points of contact to remove all varnish and corrosion.

2. Connect the wires from the explosive, one to the A terminal and one to the C or G terminal. Scrape the wires where they make contact.

3. To activate the generator wrap several turns of thin rope or wire around the drive pulley. The more turns there are the better the chance of success. Place one foot on the generator to secure it and give the cord or wire a hard pull.

FIGURE 4
Using A Bicycle Generator To Set Off Explosives

USING A BICYCLE GENERATOR TO SET OFF EXPLOSIVES

A bicycle generator that creates 6 volts and 3 watts will set off one or two No. 6 blasting caps or S95 squibs.

MATERIAL

Bicycle generator (6 volt - 3 watt or more)
Copper wire

PROCEDURE

1. Connect one wire from explosives to the generator terminal. Attach the other wire to some point on the generator case. The wire may be held against the case with the hand or wrapped around a convenient projection. Scrape the wires free of varnish or corrosion at the points of contact.

2. Activate the generator by running the drive wheel rapidly across the palm of the hand.

V. Conclusions & Future Work

Two powerful batteries, the ammonium chloride and the lye, can be rapidly constructed from materials accessible in most parts of the world. Both of these batteries have been tested and can be used to detonate explosives placed three hundred feet distant. When depolarizing materials, such as manganese dioxide, are not available, a ten cell ammonium chloride battery can be constructed which will also detonate these explosives.

Batteries made of organic materials such as urine and tea produce only low currents. A battery consisting of many more than ten of these cells will be required to reliably detonate explosives. However, the urine and tea cells, as well as the ammonium chloride cell less depolarizer, will, when combined in batteries of ten or more cells, detonate the No. 6 blasting cap.

Electrode materials such as iron (knives) and tin (cans) as well as those that are not thoroughly clean are not satisfactory for use in batteries.

The organic batteries present a very large area for future study. The study under this contract was limited to human and wasp excretions and a few natural product extracts. Excretions of other animals and other plant extracts need to be investigated for use in indigenous batteries which can give higher currents.

Six-volt automobile and bicycle generators can also be used to detonate explosives from a distance. The bicycle generator is especially useful for this purpose as it can be used repeatedly, operates at relatively low RPM and is small and easily concealed.

Section II.
Electronic Trigger Device

I. Introduction

In special warfare applications a need exists for a system to detect enemy forces approaching a perimeter defense line or to protect an area by triggering mines or similar devices when the area is intruded. Trip wire systems have been used for years to locate and disarm and, in the case of a warning device where the wire is used to initiate a signal to the sentry, the trip wire will alert the enemy intruder as well.

To improve on the basic mechanical trip wire concept, a simple electronic control circuit has been designed, fabricated, and tested to permit the electrical firing of claymore mines, explosives charges, or for triggering alarms and signal lights when an extremely thin electric wire is broken. The circuit makes use of a silicon controlled rectifier which switches the battery voltage to the detonator when it senses the change in a minute current flowing through the break wire. The low current permits the use of wire almost as thin as human hair which is difficult to detect and easily broken. In addition, the low current drawn by the circuit will permit its operation for months from the firing battery while still providing the self sterilizing feature of eventually discharging the battery to prevent detonator firing.

II. Description & Setup Procedure

DESCRIPTION OF DEVICE

The electronic triggering device consists of a controlled rectifier, resistor, capacitor and indicator lamp housed in a plastic capsule. Of the six leads emanating from the device, two are attached to a battery of 4½ to 7½ volts (e.g., BA-31 or BA-34), two leads are attached to the detonator, electric blasting cap or alarm bell and two are attached to the break wire circuit. The break wire resistance may be as high as 200 ohms permitting as much as 75 feet of AWG No. 44 wire (0.002" diameter) or 700 feet of AWG No. 34 (0.006" diameter) wire to be used in the break wire circuit. While the actual break wire should be no heavier than AWG No. 34 to permit easy breakage and make detection difficult, long lengths of heavier wire may be used between the break wire and the firing device. For example, one-half mile of blasting or field telephone wire may be used between the device and the break wire if break wire length is limited to 400 feet of No. 34 wire or 50 feet of No. 44 wire. This setup method can be useful in an alarm application where the device is coupled to a buzzer or light to indicate the presence of an enemy.

Breaking the wire switches the battery voltage across the lamp and the external firing or alarm circuit. The current carrying capacity of the device is 0.5 ampere continuous or over 5 amperes for short times such as that involved in firing a detonator. While the system is active, with the break wire intact, the current drain is quite low, in the order of 2 milliamperes. A battery such as a BA-31, 3 penlite cells in series or similar power sources will operate for 30 days or more with enough reserve to fire standard detonators such as J2 blasting cap. Since a continual current drain exists on the battery, the circuit will eventually render itself safe as the battery is exhausted. The time required for a BA-31 battery to dis-

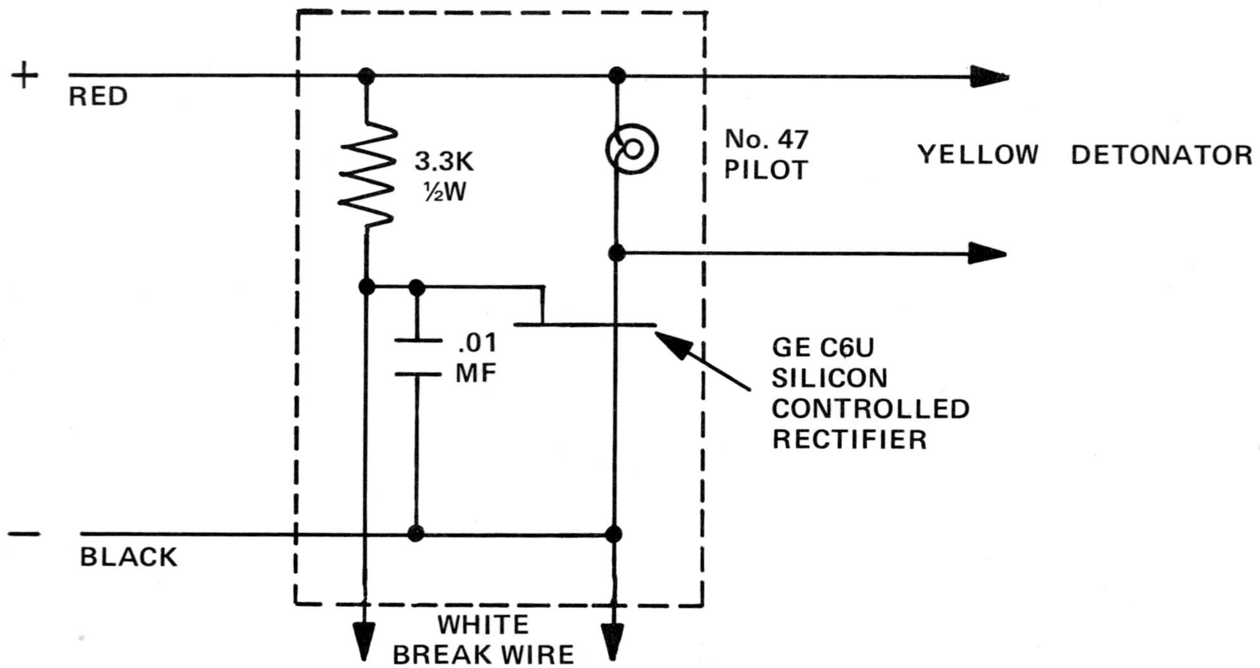

FIGURE 5
Circuit Diagram of Electronic Triggering Device

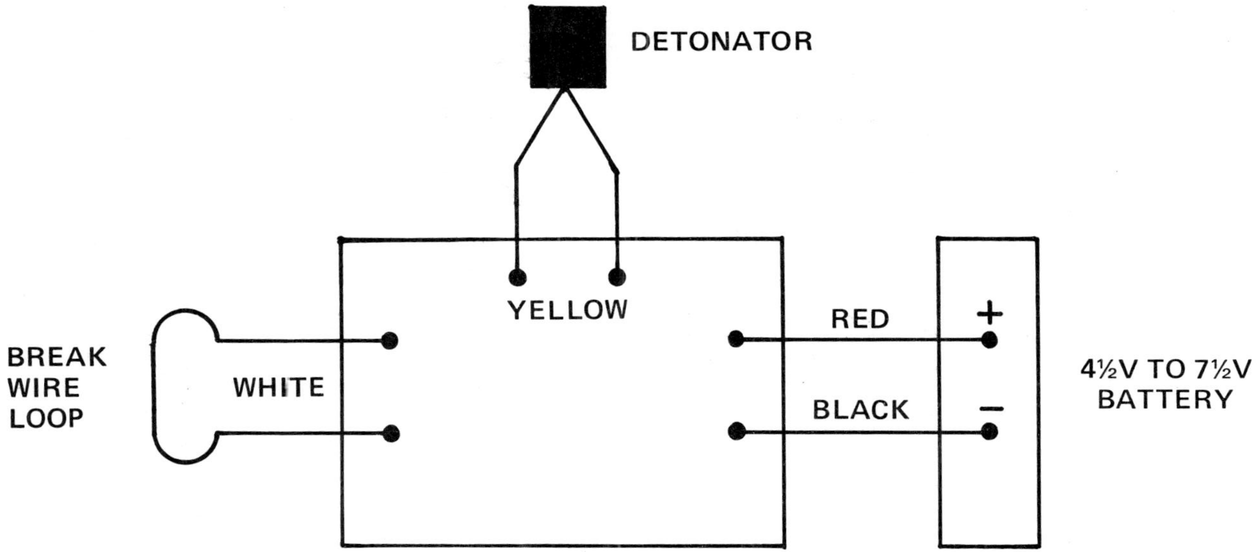

FIGURE 6
Wiring Diagram of Electronic Triggering Device

charge to the point where it will no longer fire a detonator will be between one month and one year. Larger capacity batteries such as a BA-200/U may be used for longer operating times. A circuit diagram of the device is shown in figure 5.

The parts are encapsulated in an epoxy case to provide a moisture resistant package capable of withstanding -65 to +165° F temperature extremes. Total cost of parts in small quantity lots was under $2.00. In quantity production it is estimated that the device could be manufactured for between $1.00 and $2.00 per unit.

SETUP PROCEDURE

A wiring diagram for applying the device is shown in figure 6.

The following steps should be used in setting up the device:

1. String out break wire and attach ends to white leads. (If break wire is enameled, clean ends with a knife or sandpaper.)
2. Attach battery leads (red to positive; black to negative) and observe indicator light. If light is on, the break wire is broken or improperly connected. Recheck break wire circuit and reset system by removing one battery lead and reconnecting. Repeat procedure if light still goes on.
3. With battery connected and light off, open one break wire (white) lead. The light should go on. If it does not, battery or device is defective and should be replaced.
4. Reconnect break wire circuit and reset the device by again removing one battery lead and reconnecting it. The light should now be out.
5. If light is out when battery is connected, the detonator may be attached safely to the yellow leads. When disabling the device, detach the detonator first.

III. Test Results, Conclusions & Recommendations

TEST RESULTS

Preliminary tests have been conducted on the device to determine minimum operating voltage to set off an electric detonator (J2 cap), the battery life, the maximum resistance of the firing circuit and similar variables. It was determined that a supply voltage of at least 4.5 volts was required to fire a cap reliably considering the voltage loss of .75 volt in the rectifier. Considering the firing current drawn by the gate of the rectifier an external break wire circuit resistance of 200 ohms or less was found permissible. Ten firing tests were conducted using 3 penlite cells in series as a power source, 75 feet of No. 44 wire in the break wire loop, and a J2 blasting cap as a load. All caps functioned. No temperature tests were conducted but all components are capable of operation at 200° F or greater. Tests to determine effects of RF radiation on the system, environmental tests, as well as electrical tests on a larger sample of the devices would be required to complete any development task, however.

CONCLUSIONS

The electronic triggering device described above should offer a safe, reliable and useful means of activating booby traps, alarms, mines and similar tasks in sabotage and unconventional warfare. The device could be produced with minimal development effort.

RECOMMENDATIONS

1. It is recommended that a sample lot be procured and evaluated to assure meeting user and military environmental requirements.

2. The device can be made in other current and voltage ratings with no difficulty. An investigation should be made to determine the need for a higher voltage device capable of triggering several detonators from a high voltage battery, or the need for a higher current device capable of operating a powerful klaxon horn or similar warning device.

Section III.
Gels

Gels

In the gel phase of this program gelling agents were tested for their ability to gel motor oils. The gelling agents which gave the strongest gel from the lowest concentration of additives were used in an automobile engine. After use, the engine was disassembled and examined for wear. After eighteen running hours of intermittent operation the gelled motor oil caused 0.004 inches wear on the crankshaft main bearings of the test engine.

Table V is a condensed listing of the gelling formulations tested. One of these formulations, N-coco-γ-hydroxybutyramide (Butoxyne 160)[1] was especially satisfactory. This compound was added to the automobile engine either as the presynthesized solid, or it was formed *in situ* as the product of the reaction between butyrolactone and coco amine. The latter procedure has an advantage in that the reactants are liquid and more easily and rapidly added to the engine. This formulation was tested in the automobile engine and caused the damage described above.

To gel the motor oil in an engine add one half pound of the solids N-coco-γ-hydroxybutyramide, or 0.18 pounds of butyrolactone and 0.32 pounds of coco amine (Armeen[2] 10 D) into the oil filler port. As the oil is heated by the running engine the butoxyne dissolves. When the oil cools the butoxyne comes out of solution as a crystalline gel with the oil held in the crystal interstices. When the engine is restarted the main bearings are oil starved until enough heat is generated to redissolve the N-coco-γ-hydroxybutyramide.

---

1 *Product of General Analine and Film Corporation, New York, New York*

2 *Product of Armour and Company, Chicago, Illinois*

TABLE V
OIL GELLING AGENTS

Gelling Agent	% Additives	Gel Formed	Comments
Toluene Diisocyanate Armeen[2] CD	5%-10%	No	
Toluene Diisocyanate Armeen[2] DMCD	5%-10%	No	
Toluene Diisocyanate Armeen[2] DM18D	10%	No	
Gel Bloc[4] 300	4%	Yes	Fair gel is formed when solution is cooled from 120°F
Fish Fatty Acid[5] 50% Sodium Hydroxide	9%	Yes	Moderately strong gel formed when solution is cooled from 120°F
Butoxyne[1]	4%	Yes	Moderately strong gel formed on cooling from 140°F
Phthalic Acid 50% Sodium Hydroxide	10%	No	
Linoleic Acid 50% Sodium Hydroxide	10%	No	
Diglycolic Acid 50% Sodium Hydroxide	10%	No	
Ricinoleic Acid 50% Sodium Hydroxide	10%	No	Precipitate
Tall Oil Fatty Acid 50% Sodium Hydroxide	10%	Yes	Formed a thin gel
Fumaric Acid 50% Sodium Hydroxide	10%	No	
Emery 3065-S Poly. Fatty Acid[6] 50% Sodium Hydroxide	10%	No	Precipitate
Di-Oleic Acid 150 50% Sodium Hydroxide	10%	No	Precipitate
Tartaric Acid 50% Sodium Hydroxide	10%	No	
Butyrolactone Armeen[2] 12D	5%	Yes	Moderately strong gel formed on cooling from 120°F

TABLE V (Cont'd.)

Gelling Agent	% Additives	Gel Formed	Comments
Butyrolactone Armeen[2] 10D	1.8% 3.2%	Yes	Very strong gel formed on cooling from 120°F
Butyrolactone Armeen[2] 14D	5%	Yes	Moderately strong gel formed on cooling from 120°F
Butyrolactone Armeen[2] 18D	5%	Yes	Moderately strong gel formed on cooling from 120°F
Butyrolactone Armeen[2] HPD	5%	Yes	Moderately strong gel formed on cooling from 120°F
Butyrolactone Alamine[7] 3	5%	Yes	Moderately strong gel when cooled from 120°F
Butyrolactone Triamine T	2.5%	No	
Butyrolactone Armeen[2] 0	2.5%	No	
Butyrolactone Armeen[2] TD	5%	Yes	Moderately strong gel formed on cooling from 120°F
Butyrolactone Armeen[2] 50	5%	No	
Aluminum Stearate	5%	No	
Alumagel[8]	5%	No	
Stayrite[8] No. 15	5%	No	
Colling Agent Formula B[8]	5%	No	
Gel 2X[8]	5%	No	
Stayrite No. 20[8]	5%	No	
Aluminum Octoate	5%	No	
Zinc Soap No. 26[8]	5%	No	
Zinc Stearate H[9]	5%	No	
Aluminum Stearate	5%	No	
Aluminum Palmatate	5%	No	

TABLE V (Cont'd.)

Gelling Agent	% Additives	Gel Formed	Comments
Metagel A-125[9]	5%	No	
Metagel A[9]	5%	No	
N-tert-butylacrylamide	5%	No	
Cyanogum 41[3]	5%	No	
Acrylamide	5%	No	
Polyacrylamide 50[3]	5%	No	
Polyacrylamide 75[3]	5%	No	
Polyacrylamide 100[3]	5%	No	
Nitrilotrispropionamide[3]	5%	No	
Tetrastearyl Tetronate[10]	5%	No	
Tyzor HS[10]	5%	No	
Tyzor AA[10]	5%	No	
Tyzor PB[10]	5%	No	
Armid 8[2]	5%	No	
Armid HT[2]	5%	No	
Armid C[2]	5%	No	
Armid 12[2]	5%	No	
Armid 14[2]	5%	No	
Armid 16[2]	5%	No	
Armid 18[2]	5%	No	
Dytol F-11[11]	5%	No	
Dytol E-46[11]	5%	No	

TABLE V (Cont'd.)

Gelling Agent	% Additives	Gel Formed	Comments
Dytol L-79[11]	5%	No	
Dytol L-80[11]	5%	No	
Dytol R-52[11]	5%	No	

[1]Product of General Analine and Film Corporation, New York, New York
[2]Product of Armour and Company, Chicago, Illinois
[3]Product of American Cyanamide Company, New York, New York
[4]Product of The Western Company of North America, Fort Worth, Texas
[5]Product of Humko Chemical Division, National Dairy Products Corporation, Memphis, Tennessee
[6]Product of Emery Industries, Cincinnati, Ohio
[7]Product of General Mills Corporation, Kankakee, Illinois
[8]Product of Witco Chemical Company, New York, New York
[9]Product of Metasap Chemical Company, Cedartown, Georgia
[10]Product of E. I. DuPont De Nemours and Company, Wilmington, Delaware
[11]Product of Rohm and Haas Company, Philadelphia, Pennsylvania

Gelling the water in the engine cooling system was also investigated as a means for causing automobile engine damage. About one pound three ounces of a low molecular weight polyacrylamide, AM-9[3], and about 1.5 ounces of ammonium persulfate polymerization catalyst, when added to the radiator of an automobile engine, will gel the entire cooling system. This gel does not melt upon heating.

---

3 *Product of American Cyanamide Company, New York, New York*

Section IV.
Personnel Marking

I. Discussion

The feasibility of using certain substances for the purpose of personnel marking is judged in part by the following factors:

(1) type of marking produced;
(2) residual effectiveness of the marking under adverse conditions;
(3) method of applying the marking substance.

In view of these considerations and test results, the following materials are deemed usable as personnel markers when applied as directed.

A. Silver Nitrate

Silver nitrate produces a semi-permanent, dark brown stain when it contacts the skin. Perspiration aids this reaction.

It is most effective when applied as a powder to foliage and undergrowth.

B. Fluorescent Dyes

Various fluorescent* dyes were tested, and the best results were obtained from those of an oil soluble nature. For this type of application, an oil solution has several good characteristics.

"Oil Fluorescent Green G" was the most suitable of the oil soluble fluorescent dyes tested. Its solutions have a green coloration under ordinary lighting conditions and are not extremely noticeable when applied to foliage. However, under ultraviolet light, these solutions have a bright yellow color with a characteristic glow.

**Fluorescent only in solution and with ultraviolet stimulation.*

Oil solutions adhere well to the clothing and skin. Repeated washings often fail to remove all traces of the dissolved dye even though no stain is noticeable under ordinary lighting conditions.

Very small amounts of residual fluorescent materials may be easily detected and positively identified. Traces of conventional staining substances often do not lead themselves to detection and identification.

For best results, "Oil Fluorescent Green G" should be dissolved in an oil base such as mineral oil or light motor oil. Motor oil was used in tests, and dye concentration was approximately two ounces per gallon.

Spray application to trails and foliage is suggested for effective coverage. Also, a film of the solution may be placed on the surface of canals and other shallow bodies of water which are known routes. Because of its green color, this film resembles a natural scum.

C. Fluorescent Powders

These powders have long been used as an aid in crime detection. They fluoresce under ultraviolet light while in the solid state. That is, they do not have to be dissolved in order to be effective.

Several colors of the powders are available. This allows the user to match the powder color and the color of the object to be coated.

These powders do not leave a stain when they contact the skin or clothing. They merely resemble a natural pollen or dust. Their small particle size allows them to adhere well in pores and crevices of the skin and clothing.

In tests, the powders were used alone and with sticking agents and carrying media. In most cases, the sticking agent consisted of a light oil or lubricating ester which was sprayed upon objects before application of the powder to them. In essence, these agents glued the powder to the objects and prevented appreciable loss due to wind, etc. However, they did not prevent the powder from being picked up by slight contact with the coated objects.

The carrying media were used in order to aid the residual effectiveness of the powders. One such medium, which was most effective, consisted of fifty percent kerosene and fifty percent fatty acid by volume. The powder concentration was approximately twenty grams per gallon. The resulting slurry was extremely sticky and remained on the skin and clothing in detectable amounts after repeated washings. This particular medium did not impair the fluourescent quality of the powder.

As mentioned above, these powders should be effective when used alone, with sticking agents, or with a carrying medium. Prevailing conditions should dictate the manner of application. However, when used alone, the powders must be applied in large quantities for positive results.

D. Testing

In the case of the fluorescent materials, small amounts of the powders and solutions were placed on cloth samples and on the skin. The areas of application were repeatedly washed and then checked under ultraviolet light for residual materials.

Checks with the ultraviolet source were made in darkened or semi-darkened areas. This practice is not always necessary, but visible light should be held at a minimum when possible. Visible light hampers the intensity of fluorescence of the dyes and powders.

Testing was done in part by applying various amounts of the materials to foliage and undergrowth. A subject was then instructed to pass through the general area, and his clothing, shoes and skin were checked for presence of the marker.

E. Materials and Equipment

Ultraviolet lighting - ultraviolet light source which emits long wave length radiation.

"Oil Fluorescent Green G" - distributed by Nyanza Color and Chemical Company, 109 Worth Street, New York, N. Y.

Fatty Acid - Emery 3389-R Dimer Acid, Emery Industries Incorporated, Cincinnati, Ohio.

Fluorescent powders - Sirchie Finger Print Labs, Berlin, New Jersey.

II. Conclusion

The marking substances mentioned above were very successful. However, there are many other substances which would be applicable. The area of fluorescent materials offers a number of possibilities.

Coloration was a major factor in the choice of test materials. Brightly colored formulations were avoided for obvious reasons.

No personnel marker can be effective unless contact is made with a subject. Therefore, methods of application are as important as the marker itself. Direct and positive methods should be more thoroughly investigated.

the COMBAT BOOKSHELF

DESERT PUBLICATIONS